YOUR KNOWLEDGE HAS VALUE

Bibliographic information published by the German National Library:

The German National Library lists this publication in the National Bibliography; detailed bibliographic data are available on the Internet at http://dnb.dnb.de .

Imprint:

Copyright © 2010 GRIN Verlag
Print and binding: Books on Demand GmbH, Norderstedt Germany
ISBN: 9783656853633

This book at GRIN:

https://www.grin.com/document/200917

Dina Heß

Brief introduction to the 'syntonic comma', the 'Pythagorean comma' and the 'schisma'

GRIN Verlag

GRIN - Your knowledge has value

Since its foundation in 1998, GRIN has specialized in publishing academic texts by students, college teachers and other academics as e-book and printed book. The website www.grin.com is an ideal platform for presenting term papers, final papers, scientific essays, dissertations and specialist books.

Visit us on the internet:

http://www.grin.com/

http://www.facebook.com/grincom

http://www.twitter.com/grin_com

Brief introduction to the *syntonic comma*, the *Pythagorean comma*, and the *schisma* and some methods to eliminate them

Dina Heß

Leeds, November 2010

Pythagoras is said to be the first to formulate a mathematical analysis of music, stating that 'the physical world is a material working out ... of numerical truth, and that this truth is immediately and easily apprehended ... in elementary musical consonances'[1]. He claimed that consonant musical sounds can be expressed by ratios of multiples of 2 and 3 but no greater primes. The numbers are given by the relations between lengths of vibrating strings (or frequencies). The ratio 2:1 represents an octave as a string, when struck, gives the same note as another only an octave lower if it is twice as long.

In the development of Western music, however, it became clear that Pythagoras's model was insufficient and did not always work in practice. While a singer or a good musician playing on an instrument like the violin or flute is able to adjust the intonation whilst performing, a keyboard instrument or harp prevents any ad-hoc adjustments. Thus, the existence of *commas* in a scale made it necessary to re-think the Pythagorean idea and develope systems of tuning where the ratios cannot be expressed as simple fractions of small natural numbers anymore.

In the following we shall define and explain the *syntonic comma* and the *Pythagorean comma* and discuss ways to eliminate these as done by several musicians and musical theorists throughout history. The Pythagorean system is based on the simple ratios 2 : 1 and 3 : 2 given by pure octaves and 5ths. Every interval in the diatonic scale following the Pythagorean idea

---

[1] The New Grove, 2nd Edition, *Pythagoras*

1

can be computed easily: in order to find the sum of two intervals, the ratios are multiplied; in order to find the difference, the ratios are divided[2]. For example,

$$4th = octave - 5th = \frac{2:1}{3:2} = 4:3$$

or a whole tone equals a 5th minus a 4th, hence 9 : 8. Two whole tones added together yield a major third (or 'ditone') with ratio 81 : 64.

Mathematically this corresponds to the concept of the logarithm, turning products into sums. In the second half of the 19th century Alexander Ellis introduced the concept of cents to measure frequencies, where an octave equals 1200 cents. To convert a frequency ratio as above, $r : 1$, to cents, the formula is

$$1200 \times log_2(r)$$

or in reverse, to convert n cents into a frequency ratio, the formula is

$$2^{\frac{n}{1200}} : 1.$$

For example, a (Pythagorean) perfect 5th given by the ratio 3 : 2 equals $1200 \times log_2(32)$ cents ≈ 701.955 cents.

The Pythagorean intonation was almost always referred to by musical theorists in medieval times when they talked about interval ratios[3]. But long before that the ancient Greeks already had discovered[4] that octaves and 5ths cannot be put together in an entirely satisfying way. Starting on an arbitrary note, A say, and adding twelve 5ths one ends up on $G^\times$ which is roughly the same note as A, the result of adding 7 octaves instead. In the *roughly*, however, lies the rub, because the overall interval covered by the twelve 5ths is slightly larger than that given by the seven octaves. To be exact, it exceeds the latter by 23.5 cents. This small interval is called the *Pythagorean* or *diatonic comma*[5].

For a mathematically more precise definition let us emphasise again that the Pythagorean system is based upon pure octaves and 5ths. The other ra-

---

[2]Ibid, *Pythagorean intonation*, see also: Friedrich Wilhelm Marpurg: Anfangsgruende der theoretischen Musik, Leipzig, 1757, Facsimile, New York, 1966

[3]The new Grove, *Pythagorean intonation*

[4]see also: Ll. S. Lloyd & Hugh Boyle: Intervals, Scales and Temperaments, New York, 1963, *The Perfect Fifth*

[5]The New Grove, *Temperaments*

tios can be computed as above and the resulting major scale[6] has the ratios[7]

$$(1:1), (9:8), (81:64), (4:3), (3:2), (27:16), (243:128), (2:1).$$

In this scale two successive notes are seperated either by a whole (major) tone (e.g. a 5th minus a 4th; $\frac{3:2}{4:3} = 9 : 8$) or an intervall which Benson calls a minor semitone with the ratio

$$256 : 243 = 2^8 : 3^5$$

(e.g. a 4th minus a major 3rd, i.e. two whole tones; $\frac{4:3}{81:64} = 256 : 243$). This is also called a *diesis* or *limma*. This minor semitone is not exactly half a whole tone, as two such minor semitones give the ratio

$$2^8 : 3^5 \times 2^8 : 3^5 = 2^{16} : 3^{10} \approx 9 : 8.$$

These two intervals are almost equal. If we substract them according to our rule above, we get the following ratio:

$$\frac{9 : 8}{2^{16} : 3^{10}} = 3^2 : 2^3 \times 3^{10} : 2^{16} = 3^{12} : 2^{19} \approx 1.013643...$$

which is slightly more than one ninth of a whole tone. This ratio is again a definition of the Pythagorean comma. One might think that it is possible to find numbers $m, n \in \mathbb{N}$ such that n minor semitones equal m whole tones. But such m, n cannot exist as $(9 : 8)^m$ has always an odd numerator and an even denominator whereas $(2^{16} : 3^{10})^n$ has always an odd denominator and an even numerator, no matter what m, n are. Therefore it is impossible for them to be equal.

Thus, a certain defect in the other intervals cannot be avoided if we want perfect octaves and 5ths. Something very similar happens, if we try to fit pure major 3rds into four perfect fifth. The resulting differential interval is called the *syntonic* or *Didymus's comma* and is 21.51 cents[8] or a frequency ratio of 81 : 808. The ratio between the Pythagorean comma and the syntonic comma is called the *schisma*[9] and is about 1.95 cents[10].

---

[6]for the corresponding minor scale see David Benson: Music – a mathematical offering, Cambridge, 2006

[7]Ibid

[8]Ibid, *Comma*

[9]http://en.wikipedia.org/wiki/Schisma, 08/11/2010, last update: 08/2010

[10]Dave Benson: Music – a mathematical offering

As a result, a chain of perfect 5ths always leads to imperfect 3rds and vice versa. The evolution of Western music to triad based music, starting from the Renaissance, as well as the desire to be able to modulate to different keys without having to adjust the intonation, inevitably led to the development of different tuning systems trying to create scales where both 5ths and 3rds can be used and still be almost consonant and where modulations to different keys are possible.

Throughout history many musicians, tuning experts, composers and musical theorists have tried to find solutions that would make the comma disappear (which we now know to be impossible, as shown above) or at least become insignificant. Some solutions were based on mathematical theories, others on experiments and experience. The resulting tuning systems are called temperaments. The general idea of tempered scales is to tune certain intervals slightly smaller or greater than pure in order to distribute (and thus minimise) the defect in a way that disturbes the harmony as little as possible. In his book *Tuning and temperament: a historical survey*, Barbour[11] categorises the (octave-based) scales into five groups[12], the Pythagorean system we already discussed, just intonation, mean-tone temperament, equal temperament and irregular systems[13].

The term *Just intonation* in the context of tuning keyboard instruments[14] describes a system of tuning, where most 5ths as well as most 3rds are tuned purely whereas 'some 5ths (often including D-A or else G-D) are left distastefully smaller than pure'[15]. Like the Pythagorean system just intonation leaves most intervals with frequencies in simple ratios of natural numbers. In order to attain pure 3rds and 6ths just intonation admits multiples of 5 in these ratios (the Pythagorean system did not allow multiples of primes greater than 3).

In regular *mean–tone temperament* the idea is to tune the 3rds pure and adjust the other intervals accordingly and as equal as possible[16] where a 3rd is divided into two equal whole tones. This is achieved in different ways.

---

[11] James Murray Barbour: Tuning and temperament: a historical survey, Courier Dover Publications, 2004

[12] see also: David Benson: Music – a mathematical offering

[13] The New Grove, *Temperaments*

[14] The New Grove gives a second concept under the same name, describing the method of adjusting intonation ad-hoc in order to attain pure intervals like explained earlier

[15] The New Grove, *Just intonation*

[16] Dave Benson, Music – a mathematical offering

The most common version of this is the *1/4–comma–mean–tone tempera-ment*, where the fifth are tuned one quarter of a syntonic comma smaller than pure. Thus the 'problem of the syntonic comma' is more or less taken care of and modulation to other keys is possible. But still not within the entire circle of 5ths as the Pythagorean comma still exists, producing an interval known as the *wolf 5th* between, for instance, $C\sharp$ and $A\flat$ which is noticably not euphoneous. Other temperaments used in Baroque- and Re-naissance tuning, where 3rds are tuned slightly smaller (or larger) than pure and divided into two equal whole tones were also called mean–tone temper-aments. For example, the 2/9–comma–mean–tone temperament where the 3rds are tuned 1/9 larger than pure.

The idea of *equal temperament*, as it is usually used today, already came up in the 16th century and probably earlier. The idea is to distribute twelve semitones equally into an octave, the latter being the only pure interval left. Today these semitones equal 100 cents where Vincenzo Galileo suggested a ratio of 18 : 17 ($\approx$ 99 cents) as an approximation for the tuning of semitones on a lute in 1588[17]. Later, 17th-century composer Girolamo Frescobaldi endorsed the equal temperament for keyboard instruments and according to his contemporary Marin Mersenne[18], the engineer Jean Gallé used it for tuning organs and spinets. However it was not until the 19th century that equal temperament became the common system of tuning for keyboard and fretted instruments. This is probably due to the fact that other tunings resulted in different characters for different keys and the fact that some sound better for certain instruments or timbres than others. Equal temperament produces a pleasant average. This is best depicted in the following quotation by 18th-century musical theorist Neidhard:

> 'Most people do not find in this tuning that which they seek. It lacks, they say, variety in the beating of the major 3rds and, consequently, a heightening of emotion...Yet if oboes, flutes & the like, and also violins, lutes, gambas & the rest, were all arranged in this same [tuning], then the inevitable church- and chamber-pitch would blend together throughout in the purest [way]...Thus equal temperament brings with it its comfort and discomfort, like blessed matrimony...'[19].

---

[17]The New Grove, *Temperaments*
[18]Ibid
[19]Ibid

Finally, the term *irregular temperaments, circulating–* or *well–tempered* systems refer to tuning systems of mostly the 16th, 17th and 18th century where the notes near to the wolf 5th in a mean–tempered scale are adjusted such that the wolf 5th is smoothed out. A rather prominent example is the tuning system introduced by Andreas Werckmeister (1691)[20] known as *Werckmeister III*, where the Pythagorean comma is distributed equally on certain 5ths in the scale. Musical theorists are convinced[21] that the tuning, Johann Sebastian Bach applied when composing the 48 preludes and fugues in *Das wohltemperirte Clavier* was a well–tempered or irregular tempered system rather than equal temperament as he was an advocate of the associating of a character to different keys[22]. This notion, however, is implausible in a system where every interval has a certain frequency independent from key or pitch[23].

Many other systems have been developed throughout the centuries to deal with the commas and still be true to perfect intervals and ratios to a certain degree. Many systems employ mathematical methods as well as consideration for the musical and practical applications. Some systems require new instruments[24], and some instruments require new (or rather old) temperaments. Every system has its own advantages and disadvantages and many composers have written music that works the best in a certain temperament. The Pythagorean and syntonic comma have always been one of the the most immediate (and obvious) links between mathematics and music and the dealing with them has certainly inspired countless musicians and mathematicians of all times and motivated many scientific discoveries in the past, beginning, of course, with Pythagoras.

---

[20]Andreas Werckmeister is also said to be the first to draw up a contract making the tuner tune the organ in the *Quedlingburger Dom* in well–tempered intonation as proposed in the afterword of: Andreas Werckmeister: Orgel Probe, 1698, Faksimile Nachdruck, Bärenreiter

[21]The New Grove, *The well–tempered clavier*

[22]see also Neidhard's quotation above

[23]see also: George Stauffer & ErnestMay (editors): J.S. Bach as Organist - His Instruments, Music and Performance Practices, Indiana University Press 1986, Harald Vogel, John Brombaugh

[24]This becomes obvious once we look at scales where an octave is devided into more than 12 semitones like, for instance, in 31–tone equal temperament

YOUR KNOWLEDGE HAS VALUE